Thorium:
The Eighth Element

BRIAN BASHAM,
EQUITY DEVELOPMENT

ADVFN BOOKS

Equity Development contacts:
Brian Basham
0207 065 2690

Andy Edmond
0207 065 2691

EQUITY
DEVELOPMENT

andy@equitydevelopment.co.uk

Please refer to the important disclosures shown on the back page and note that post MiFID this information is categorized as Marketing Material

CONTENTS

THE EIGHTH ELEMENT

The Eighth Element is named after Lord Browne's recently published book 'Seven Elements That Have Changed the World'.

The elements in Lord Browne's list are iron, carbon, gold, silver, uranium, titanium and silicon. Uranium is there, not only as a global nuclear fuel, but also as the source of plutonium 239 which is the primary fissile isotope used for the production of nuclear weapons.

This paper argues that thorium deserves the title 'The Eighth Element' as a better, safer, cheaper and more prolific replacement for uranium.

Thorium mixed with plutonium to create a nuclear fuel, destroys plutonium thus creating crucial resistance to nuclear weapons proliferation and the terrifying risk of the acquisition of plutonium by terrorist groups.

Moreover, because plutonium is destroyed, the cost of decommissioning a reactor is, experts opine, likely to be reduced by circa 90 percent.

According to the Public Accounts Committee report earlier this year the cost of cleaning up the Sellafield nuclear waste site in Cumbria is a staggering £67.5bn and costs continue to mount at the rate of £1.5bn per annum, with no end in sight.

The Public Accounts Committee report suggested that successive governments have failed to "get to grips" with the hordes of waste stored at the site.

Today the cost of decommissioning a uranium fuelled reactor greatly inflates the cost of construction. Most of this cost is allied to the storing and disposal of the plutonium produced by such reactors. By comparison the cost of decommissioning a thorium fuelled reactor is negligible and the construction of a Th-100 may, on preliminary estimates, prove to be as low as c. $250m.

Switching to thorium would be popular with the electorate because it would increase safety, vastly reduce the huge burden on future

generations of decommissioning costs and provide access to a fuel source that is many hundreds of times more accessible than uranium. Forty years ago, thorium was actively being developed in the US, Britain and Germany as a nuclear fuel but, at the height of the Cold War, that work was abandoned for a variety of reasons, none of which reflected upon thorium's efficacy as a nuclear fuel. Perhaps the dominant of those reasons was that the thorium fuel cycle does not produce Pu239, the favoured isotope for nuclear weapons' production.

As noted, a thorium fuelled reactor can be used to eliminate troubling nuclear waste by burning a thorium fuel that is mixed with plutonium.

Such reactors could be used to destroy the circa 1,850 tons of plutonium that is held by 35 countries and which includes an estimated 70 tons in nuclear warheads. Ed Fei, senior policy adviser at the US Energy Department Global Threat Reduction Initiative has spoken of the need to *'clean up leftovers of large nuclear supplies to prevent terrorists from obtaining them'.*

At 112 tons, the UK holds one of the world's largest plutonium stocks. Dr Tim Fox, Head of Energy and Environment of the Institute of Mechanical Engineers has said: *'There are estimates that it currently costs the UK taxpayers £80m a year to store safely the country's civil held plutonium'.*

Thorium is also four times more prevalent in the earth's crust than uranium and it is a much more 'efficient' fuel. Thorium can be used in its 'raw' state as a fuel, whilst uranium requires extensive processing and enriching. All of, say, one ton of thorium can be used whilst only a fraction of a ton of uranium actually serves as the fuel.

In fact it takes 40 tons of naturally occurring uranium to produce the same energy as one ton of naturally occurring thorium; consequently it can be said that the world's total supply of thorium could produce energy to an amount 160 times that of the total world supply of uranium.

With all these attributes why it is that thorium has not taken off?

As one of the world's best known oilmen with a vast knowledge of energy production Lord Browne is well aware of the power of the potential importance of thorium and during his career as chief executive of BP, he visited thorium fuelled nuclear projects in China, Germany, and India.

He has the answer to the mystery of thorium's neglect and it is remarkably abrupt. *'Vested interests'* he says. The industrial power structure of the world is against change – even with all those wonderful attributes.

Now, however, the time has come for change, in the wake of the Fukushima Daiichi nuclear disaster, in the face of a looming energy crisis, with climate change fears to the fore and with the constant threat of terrorists getting their hands on plutonium.

This paper argues that it is time for the UK, and other governments, to develop a safe thorium based nuclear strategy. A strategy that will, over time, eliminate the UK's vast plutonium stockpile, that will accelerate the UK's nuclear programme and will make a major contribution to solving the UK's energy problems without the danger either of nuclear meltdown or of creating more plutonium.

Thorium Facts

- 1 ton of thorium should generate as much energy as 3,400,000 tonnes of coal.

- Thorium reactors destroy plutonium, so their decommissioning costs are fraction of a uranium fuelled reactor.

- STL and Thor Energy are companies committed to delivering thorium energy to the world.

- Thor Energy's thorium fuel has won support from the Norwegian Government and interest from around the world.

- New reactor designs have attracted interest from the US Department of Energy.

- In April 2013, Thor Energy successfully installed six thorium fuel rods into the Halden Reactor.

- With access to thorium itself, plus design of a new modular reactor, STL is effectively an option on take-up of this energy source.

Introduction

We have been asked to write this paper as a natural update to our previous written work on this subject.

In November 2011 we published a non-commissioned paper called '*A Chance to Rethink UK Energy Policy*'. We were prompted to do so by reports from the Weinberg Foundation meetings held at the House of Lords and sponsored there by Baroness (Bryony) Worthington, the Foundation's patron.

The Weinberg Foundation is a UK based not-for-profit organisation dedicated to advancing the research, development and deployment of safe, clean and affordable nuclear energy technologies to combat climate change and to underpin sustainable development for the world.

It was named in honour of Alvin Martin Weinberg, who died in 2006, the nuclear scientist who came to pioneer peaceful nuclear technology. The Weinberg Foundation has been sponsored by descendants of Alvin Weinberg.

Weinberg was sacked by the Nixon administration from Oak Ridge National Laboratory in 1973 after 18 years as the lab director because he continued to advocate increased nuclear safety in molten salt reactors instead of the administration's chosen Liquid Metal Fast Breeder Reactor (LMFBR). He built an experimental molten salt reactor (MSR).

The Weinberg Foundation has continued to promote the thorium fuelled Molten Salt Reactor (MSR). However, despite widespread support for the concept, an MSR has yet to be built.

As a result of our '*Rethink*' paper we were approached by a nuclear scientist, Julian Kelly, an Australian who is a consultant to Thor Energy in Norway.

A short description of Thor Energy appears later in this document but in summary, it has been part financed by the Norwegian Government to develop the uses of thorium as a nuclear fuel.

Thor Energy is now engaged in a five year project to irradiate thorium, plutonium mixed pellets in the Halden reactor in Norway, which we have visited. The irradiation and qualification process began in April this year and Thor Energy reports widespread interest in the project already from around the world.

It is an exciting time for the Thor Energy team because as they demonstrate success they will be able to penetrate the market for using their thorium, plutonium fuel in the circa 360 light water reactors (LWRs) that are in world today. In the process, they will begin to burn up plutonium stocks.

It was through our association with Thor Energy that we became aware of STL, which has acquired a 15 per cent interest in Thor Energy. STL appears, currently, to be the only way for investors to secure a direct interest in the future of thorium.

STL is the owner of the rights to what will become one of the world's largest stocks of high grade thorium. It is also the developer of the Th-100 thorium based Modular Pebble Bed Reactor (MPBR). It is the combination of those factors that appeals.

We were attracted to the Th-100 for eight principal reasons:

- It is safe! It cannot overheat and go into meltdown.

- Using thorium as a fuel mixed with plutonium it burns the plutonium to destruction.

- It is based upon tried and tested pebble bed reactors that have been in operation in Germany and China and in development in South Africa.

- The Th-100 reactor is ready for commercialisation, with a projected seven year delivery time that is in stark contrast to the decades needed for other nuclear alternatives.

- It does not generate the hydrogen gas which played such a large part in the Fukushima disaster.

- It is modular and the components can be transported by road.

- It uses many commonly available components.

- STL has learned from the operation of the South African project and has recruited a top team of people from the South African reactor programme.

We shall look more closely at the companies later in the report, but first let us remind readers of the characteristics of the element itself:

Thorium

The primary benefits of utilizing thorium as a nuclear fuel are safe, clean energy production, enhanced proliferation resistance, and potential fuel cycle cost savings.

The thorium-based fuel has improved thermo-physical properties compared to uranium fuels, which provide more reliable fuel performance during reactor operation. The lower operating temperature and higher thermal conductivity of thorium fuel leads to increased safety margins during off-normal and accident scenarios.

Nuclear proliferation refers to the spread of special nuclear material (material that can be used to make a nuclear weapon) to nations or organizations that are not recognized as Weapons States under the Treaty on the Non-Proliferation of Nuclear Weapons.

All nuclear fuels and nuclear reactors have an inherent nuclear proliferation risk. In order to have a sustainable nuclear chain reaction (necessary for power production) one must use fissionable material. A nuclear weapon uses these same materials in much higher concentrations to achieve an uncontrolled nuclear chain reaction.

There are several methods to enhance the proliferation resistance of a nuclear fuel cycle which are classified as intrinsic and extrinsic barriers.

Intrinsic barriers are inherent to the physical properties of the fuel and the technical capabilities required to produce weapons from it. Extrinsic barriers (also called institutional barriers) include a wide array of nuclear safeguards such as material control and accountability and numerous international agreements.

Thorium-based nuclear fuel provides enhanced proliferation resistance through several intrinsic barriers (related to the properties of the fuel and the technology required to produce a weapon).

Plutonium is a by-product of the nuclear fission process. The isotope Pu-239 is a weapons-useable isotope and is the primary concern with regards to making a plutonium weapon. Other isotopes of plutonium actually hinder the ability to make a successful plutonium weapon (e.g. Pu-238 & Pu-240).

The thorium-based fuel used by STL and Thor Energy employs plutonium as the driver for the thorium plutonium mixed fuel. The plutonium is thereby consumed and no new plutonium is produced.

The fuel is used in a "once-through" fuel cycle, for which reprocessing is not required on the used thorium fuel to take advantage of the energy content of the U-233. This allows it to utilize a significant amount of the generated U-233 before the fuel leaves the reactor, reducing the residual energy value of the used fuel.

Other intrinsic barriers exist to increase proliferation resistance: the radiation-dose-related problems associated with handling U-233 are considerable and more complex than simply enriching natural uranium for the purpose of bomb making.

Thorium is a naturally occurring isotope and decays much slower than the other long-lived isotopes in used fuel and the radiation it gives off (so-called α-particles) travels only a few centimetres in air and can be stopped by human skin or a sheet of paper.

Thorium-dioxide is more stable than uranium-dioxide especially with respect to the fact that thorium does not dissolve in water whereas uranium does. Therefore, the likelihood of ingesting material from thorium-based fuels is lower than conventional uranium fuel because the thorium fuel would not disperse through water in the event of a storage system failure.

There is no international or standard classification for thorium resources and identified Th resources do not have the same meaning in terms of classification as identified U resources. Thorium is not a primary exploration target and resources are estimated in relation to uranium and rare earth resources.

Estimated World Thorium Resources

Country	Tonnes	% of total
India	846,000	16
Turkey	744,000	14
Brazil	606,000	11
Australia	521,000	10
USA	434,000	8
Egypt	380,000	7
Norway	320,000	6
Venezuela	300,000	6
Canada	172,000	3
Russia	155,000	3
South Africa	148,000	3
China	100,000	2
Greenland	86,000	2
Finland	60,000	1
Sweden	50,000	1
Kazakhstan	50,000	1
Other countries	413,000	8
World total	**5,385,000**	

Source: World Nuclear Association

STL

Steenkampskraal Thorium Limited is a South African company located near Johannesburg. It is developing thorium as a clean safe energy source. STL's objective is to exploit the thorium value chain from the mine through to the reactor.

STL has the rights to a 'free carry' on the thorium produced by the Great Western Minerals Group (GWMG) as a by-product of that company's rare earth mining activities in **Steenkampskraal.** Mining is set to begin in 2015 and should deliver thorium to STL at the rate of 600 tons a year. The present size of the measured resource indicates that STL might eventually own something like 10,000 tons of *high grade* thorium.

Their Th-100 nuclear reactor consists of a 100 MW high temperature, helium cooled, pebble bed reactor. It features a Once-Through-Then-Out (OTTO) Thorium fuel cycle, thereby simplifying the layout, whilst simultaneously enhancing the proliferation resistance characteristics by performing an in-situ burn up of driver material, namely plutonium isotopes.

The Th-100 produces steam as its energy product and is therefore versatile in its application. Steam can be used for producing power via a steam turbine (35MW), or it can be used for process heat in petrochemical plants, oil refineries and many other applications.

Development of new reactors has attracted close attention from a number of government agencies, especially in the US including the Department of Energy and the Nuclear Regulatory Commission.

The Department of Energy has set aside $1 billion to fund research into small modular reactors. Firstly, a $400 million grant has been given to Babcock and Wilcox and the Tennessee Valley Authority to build a small modular light water reactor. Of course, this reactor is not meltdown proof and it will generate yet more plutonium. Secondly, $450 million has been set aside to fund what is defined as Generation 3+ and Generation 4 reactors. Generation 3+ are LWRs with much enhanced safety features but the

Generation 4 definition is more demanding in that it has to be 'meltdown proof'.

There are three reactors in competition for these funds one of which is small pebble bed reactor that could be an ideal consumer of a thorium-based fuel.

Other particularly attractive features of the Th-100 are that: using thorium as a fuel mixed as plutonium, it burns plutonium to destruction; and that it does not generate the hydrogen gas that played such a large part in the Fukushima disaster.

The estimated economics of construction and production are at the moment based on incomplete information. However, at a 'ball-park' cost of c. $250m per reactor one could see power generation at 12c / kWh as potentially achievable.

In summary, therefore, the Th-100:

- Is meltdown proof. Its fully ceramic fuel elements cannot melt, even in the event of total loss of active cooling. Pebble bed reactors are the only meltdown proof reactors available today.

- Uses coated thorium fuel particles (TRISO) effectively retaining the fission products within the fuel and allowing for very high burn-up of the fuel.

- Reactor core can tolerate a loss of forced cooling event and a total loss of the decay heat removal capability. Passive decay heat removal is possible and fuel temperatures stay below admissible values. Therefore, the fission products remain inside the fuel particles even in extreme accidents.

- Has a very strong negative temperature coefficient that contributes to its excellent inherent safety characteristics.

- Employs helium as coolant, which is both chemically and radiologically inert and does not influence the neutron balance. It allows for very high coolant temperatures during normal operation.

- Reactor core has a low power density, providing a very robust design with high heat capacity which renders the reactor thermally inert during all operational and control procedures.

- Efficient retention of fission products in the coated particle fuel in normal operation allows for a clean helium circuit; resulting in low levels of contamination of the coolant gas, low release of radioactivity and extremely low radiation dose values to the operation staff and the elimination of the risk of catastrophic release to the environment.

- Fission product release is protected by multiple independent barriers.

The pebble bed design and fuel provide unique safety features which ensure that fission products are retained within a number of independent barriers, as illustrated in the next table:

UNIQUE SAFETY FEATURES		
Barrier (containment)	Pebble bed reactor	Traditional water cooled reactors
First barrier	Triso coated fuel particles	Zircaloy tubes or cladding
Effect of first barrier failure on the second barrier during LOCA*	Failure of fuel particles has no effect on reactor pressure vessel. Failed particles would release small amounts of fission products into the primary circuit but would be contained within the pressure vessel.	Failure of zircaloy fuel pins releases large amounts of fission product and melted fuel into the core. This molten fuel collects in the bottom of the pressure vessel and can melt through the pressure vessel steel resulting in failure of the second barrier.
Second barrier	Reactor pressure vessel	Reactor pressure vessel
Failure of the pressure vessel	Failure of the pressure vessel would release helium into the reactor cavity. Helium is radiologically and chemically inert and will therefore have no effect on the containment building. Helium being in gaseous form means there is no phase change which results in severe pressure buildup due to phase change in the reactor cavity. Helium can be vented into atmosphere passing it through carbon filters to remove possible contaminated dust particles before releasing uncontaminated helium to atmosphere.	Failure of the pressure vessel would release steam and hydrogen into the reactor cavity. This is an explosive mixture and due to the water-steam phase change results in severe pressure buildup in the containment building. This mixture cannot be vented as it is a highly contaminated mixture of gasses. At Fukushima this explosive mixture exploded and resulted in the final barrier (i.e. the containment) being breached and thus large amounts of contaminated gas and water were released into the environment.
Third barrier	Reactor containment building	Reactor containment building
*LOCA – Loss of Coolant Accident – this is what occurred at Fukushima		

Source: STL

Diverse Usage

The Th-100 power plant can be configured to cater for a variety of applications:

- Electricity generation using a standard off the shelf steam turbine generator,

- Process steam for use in chemical plants, paper mills or petrochemical plants,

- Heat for desalination plants,

- CO_2 free steam for lifting oil from oil sands,

- Power pack for off-shore oil platforms,

- Off-grid distributed power on islands and remote areas,

- Power plants for large electricity users such as smelters, cement plants, refineries and other applications.

Thor Energy

STL owns a 15 per cent share in the Norwegian company, Thor Energy, which is a subsidiary of the Scatec group of renewable energy enterprises based in Oslo, Norway.

Thor Energy is developing the use of Thorium in traditional LWR nuclear plants and has therefore started with an extensive thorium-mixed oxide fuel test program and for this it has received project funding from the Norwegian Government. **The fuel is currently being irradiated in Norway's Halden reactor, as part of the fuel qualification program.**

Thor Energy has built an impressive international consortium by which to realize this major undertaking. Members of the consortium include Westinghouse, which is now owned by Toshiba, the Finnish utility company, Fortum, and the UK's National Nuclear Laboratory (NNL) based at Sellafield. The members will collectively steer the irradiation experiment, co-fund the undertaking and share all resulting data. The consortium remains open to all interested parties and Thor Energy has received expressions of interest from around the world.

Thorium fuels in the broadest sense can provide avenues to improve the credentials for nuclear energy by:

- Achieving more sustainable energy generation in which mined nuclear material is used more effectively. This draws on the possibility for high-conversion or even breeding of fissile U233 from thorium fuels in thermal reactors.

- Employing fuels that generate smaller problematic waste streams, and that can also transmute (destroy) actinide components, such as plutonium, in current-generation thermal reactor systems.

Thor Energy's technology development activities are undertaken with the vision that thorium-plutonium MOX analogue fuels will be an attractive option for both light water reactor operators and nuclear energy policy makers alike.

As ever, nuclear power generation is less expensive and more reliable with longer lasting plants compared with wind and solar generation.

Thorium – in summary

Since our last note drawing attention to the benefits of using thorium as a fuel, we have benefitted from further researching the associated technology and the companies well advanced in making plans a reality.

We have produced this updated report to help further raise awareness amongst society's decision takers of a necessary course to be pursued, and to help the businesses themselves receive the support necessary to deliver this global energy solution.

ABOUT EQUITY DEVELOPMENT

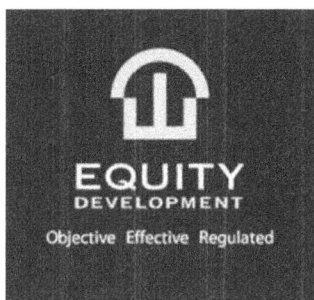

Equity Development enables companies to become better understood and supported by investors. Since our launch in 1996 we have consistently focused on helping our clients improve their communication and relationships with both existing and potential shareholders. Our clients have come from a wide variety of sectors and domiciles, are both private and quoted and range in size from micro-cap to FTSE mid-250.

Our team brings together a range of committed professionals proven at the highest level in diverse financial environments: from smaller company corporate broking in the UK to global investment banking. Focused on delivering a premium service for a select number of companies, our collective experience brings additional value to clients through advice when they need clear opinions and typically leads to client relationships that last many years.

Our analysts are all drawn from experienced City backgrounds (no teenage scribblers!) who have the freedom to commit time to producing detailed, insightful research. We stand by the quality of our product, ensuring that we differentiate ourselves from other commentators with thought provoking coverage, not 'research by numbers'.

www.equitydevelopment.co.uk

Equity Development Sales Contacts

Head of Corporate
Gilbert Ellacombe
Direct: 0207 065 2698
Tel: 0207 065 2690
email: gilbert@equitydevelopment.co.uk

Investor Access
Hannah Crowe
Direct: 0207 065 2692
Tel: 0207 065 2690
email: hannah@equitydevelopment.co.uk

Senior Analyst Contacts

Consumer, Support Service
Ben Maitland
ben@equitydevelopment.co.uk

Natural Resources
Conor Fahy
conor@equitydevelopment.co.uk

Technology
Denis Gross
denis@equitydevelopment.co.uk

Financials
John Borgars
johnb@equitydevelopment.co.uk

Special Situations
Paul Hill
paul.hill@equitydevelopment.co.uk

Transport
Matthew O'Keeffe
matthew@equitydevelopment.co.uk

Pharma / Biotech

Lorenza Castellon

lorenza@equitydevelopment.co.uk

TMT

John Walter

johnw@equitydevelopment.co.uk

Strategy

Andy Hartwill

andyh@equitydevelopment.co.uk

ABOUT ADVFN

ADVFN
www.advfn.com

ADVFN is the world's leading financial market website.

With full real-time coverage of the London Stock Exchange, the euronext-liffe, PLUS Markets, NASDAQ, Amex and NYSE, including FTSE, Dow and S&P indices, ADVFN provides professional quality information to the person on the street.

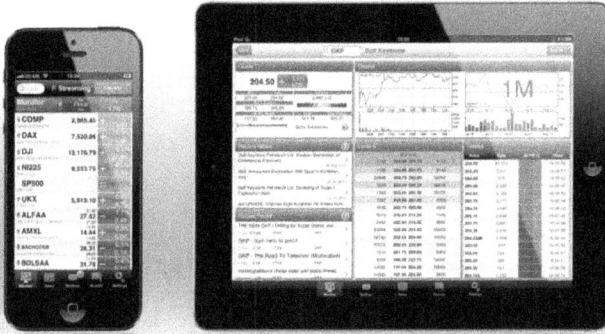

www.ingramcontent.com/pod-product-compliance
Lightning Source LLC
Chambersburg PA
CBHW060514200326
41520CB00017B/5042